MW01612372

Amanda Neel

Doodle Dogs

Coloring Book For Adults

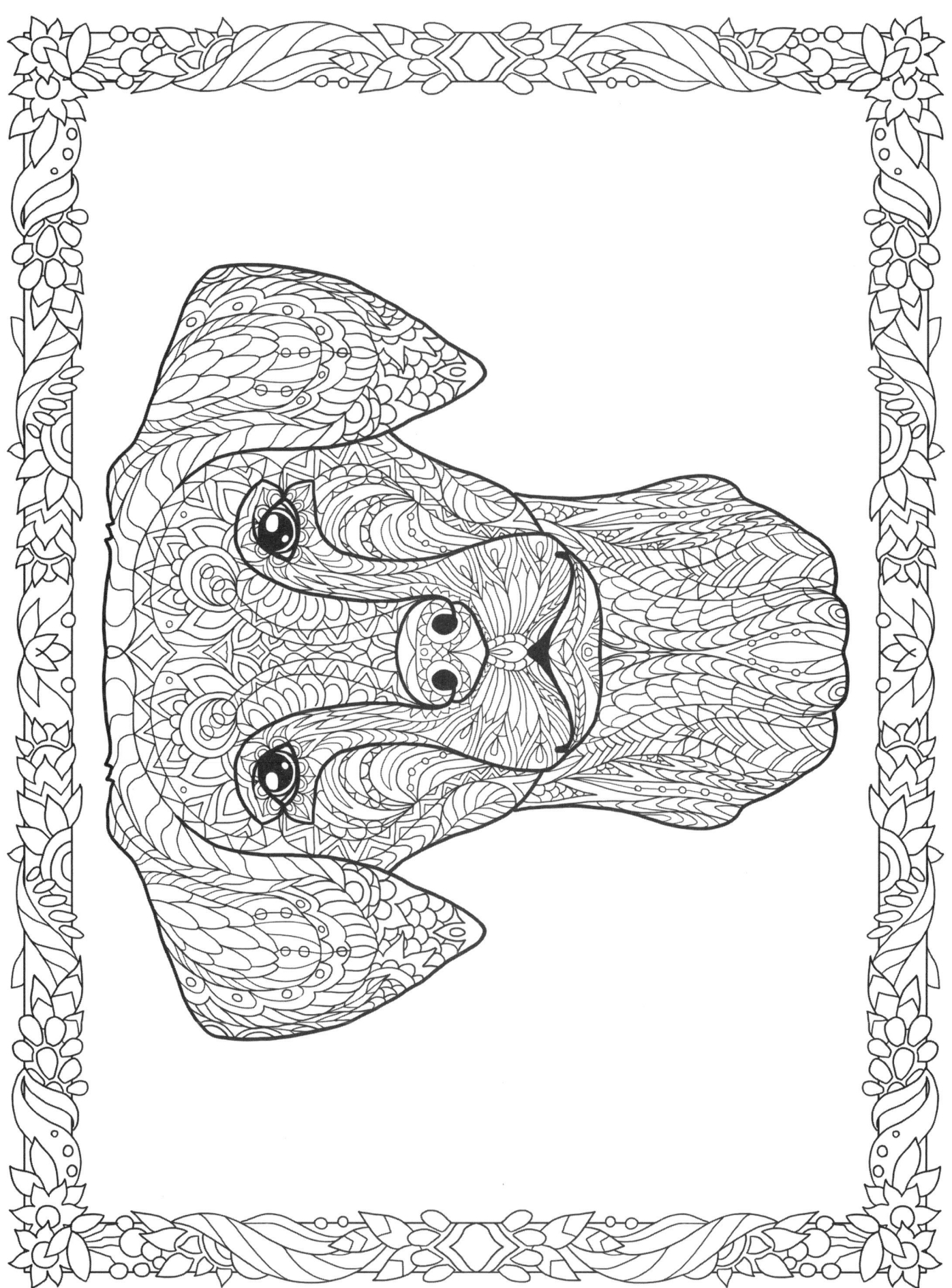

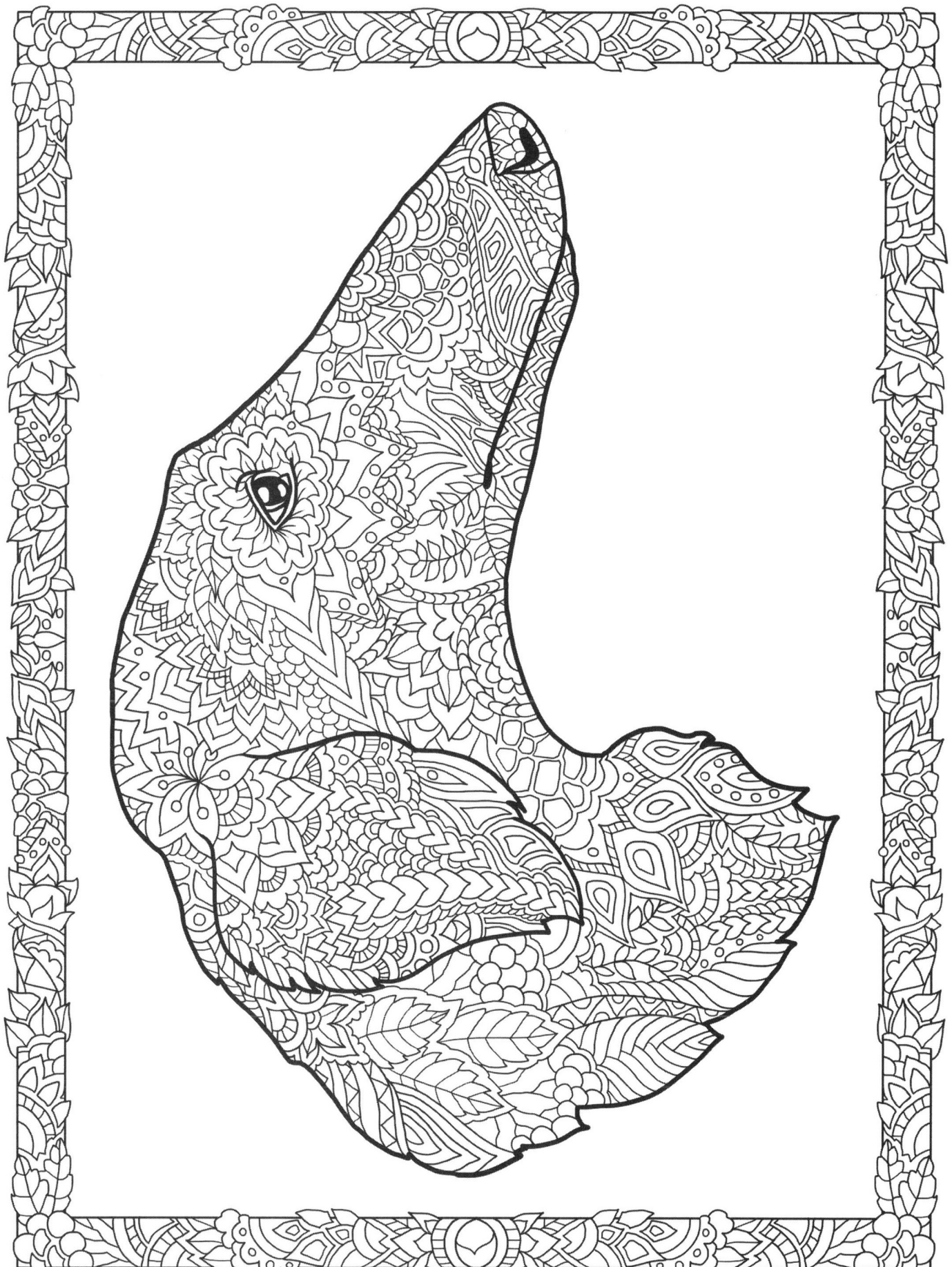

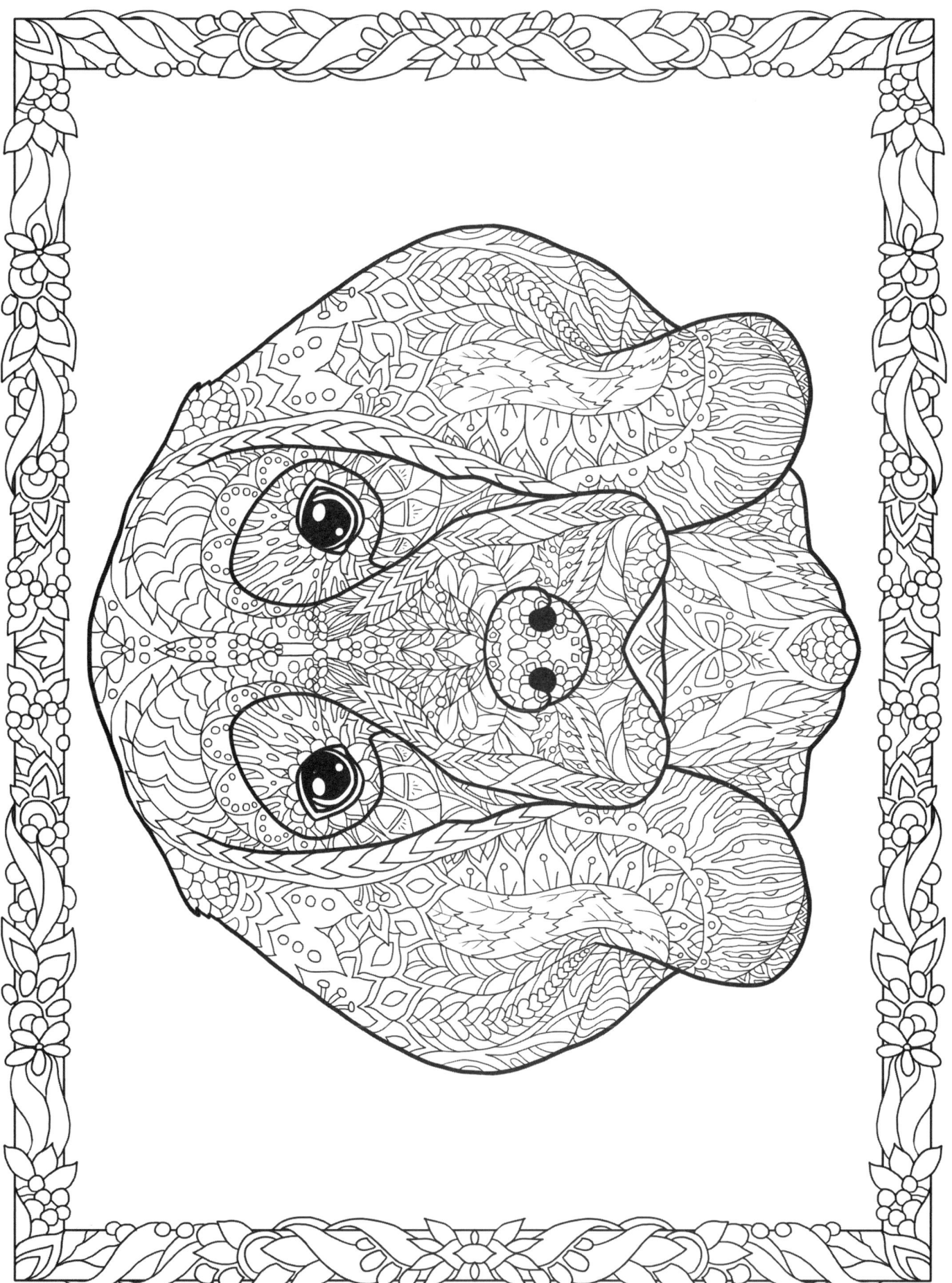

A SPECIAL GIFT FOR YOU!

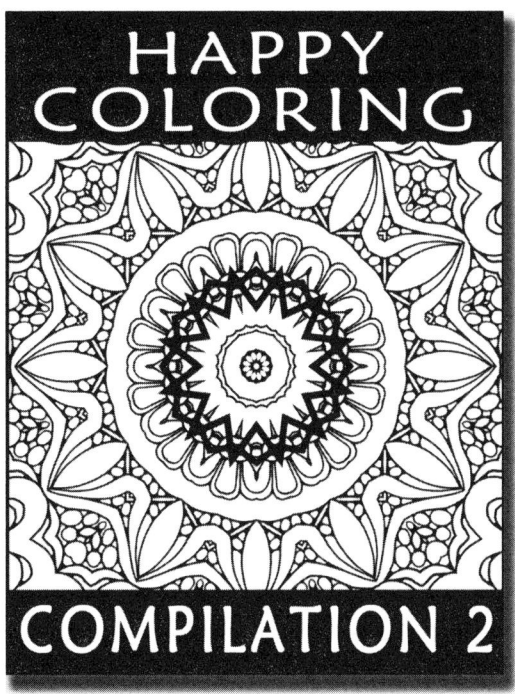

HAPPY COLORING - COMPILATION 2 is a Free Digital Book.

This coloring book features 13 illustrations selected from HAPPY COLORING books series! Illustrations in this book include pieces with all levels of difficulty, from very simple coloring to intricate. Download the PDF files and print them on 8.5"x11" quality paper.

To download the PDF files simply register here:

www.happycoloring.com/freecompilation2

Enjoy!

Happy Coloring

Doodle Dogs

Coloring Book for Adults

In this coloring book there are thirty doodle style illustrations representing dogs in a range of complexity from beginner to expert-level. I hope you enjoy this book and get positive emotions to color the images. If you like this book, please take a moment to post a review on www.amazon.com

Happy Coloring!

Amanda

About Amanda Neel

Amanda is an illustrator with a degree in Graphic Design. Her favorite subjects are related to nature. Her passion is to draw cats, dogs, horses and other animals.

Doodle Dogs Coloring Book for Adults

ISBN-13: 978-1533625649
ISBN-10: 1533625646

Please feel free to contact us if you have any questions or comments: altispublishing@hotmail.com

From the same book series:

☐ Lovely Cats - Coloring Book for Adults   ISBN-13: 978-1518706127

☐ Beautiful Horses - Coloring Book for Adults   ISBN-13: 978-1519277169

☐ Lovely Dogs - Coloring Book for Adults   ISBN-13: 978-1522921714

☐ Amazing Swirls – Coloring Book for Adults   ISBN-13: 978-1519703644

☐ Animals – Coloring Book for Adults   ISBN-13: 978-1519399052

☐ Butterflies – Coloring Book for Adults   ISBN-13: 978-1530174171

Visit our websites:

www.happycoloring.com
www.pinterest.com/happycoloring
www.facebook.com/happycoloringbooks